MW01462657

This Notebook Belongs to

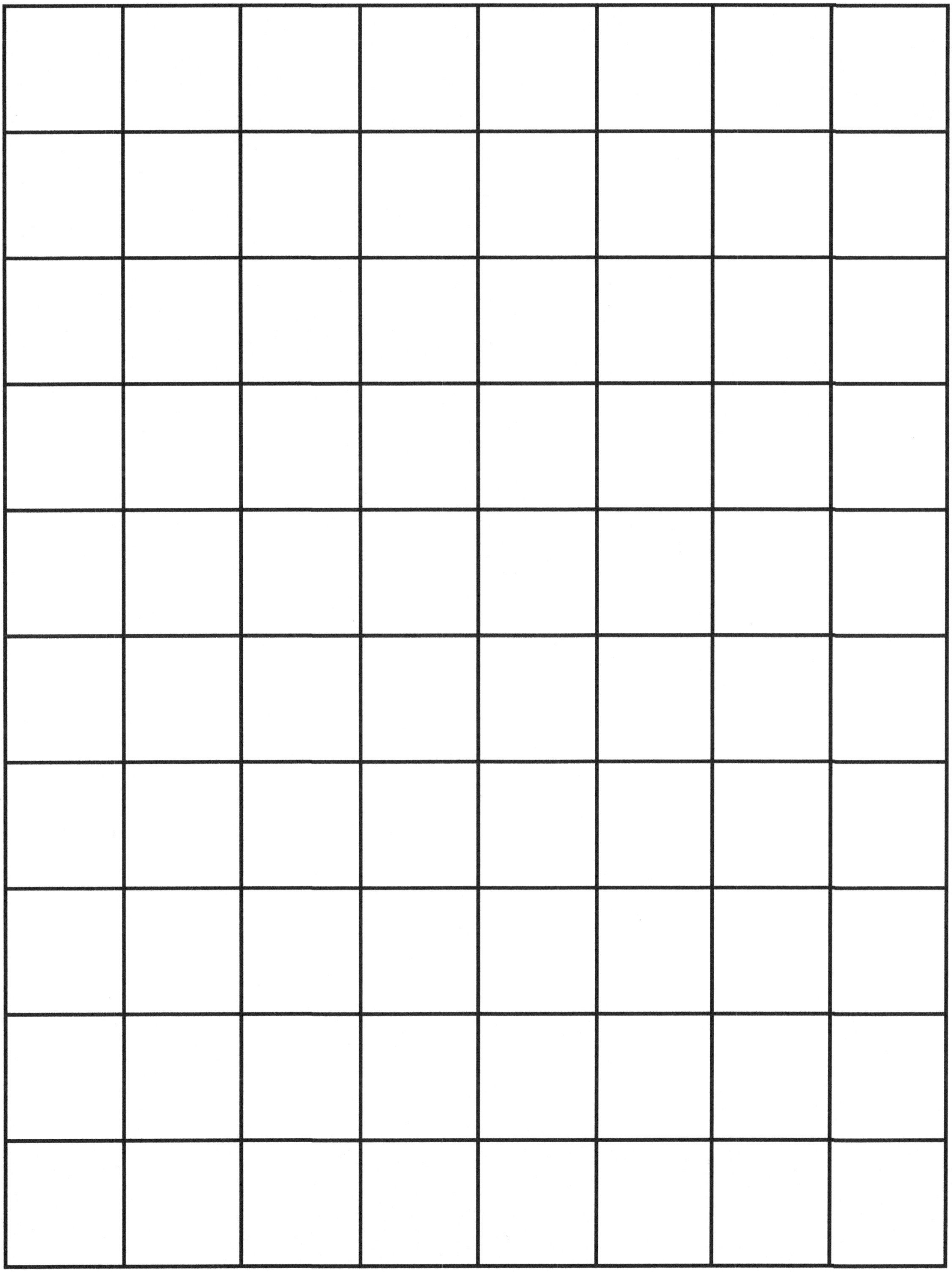

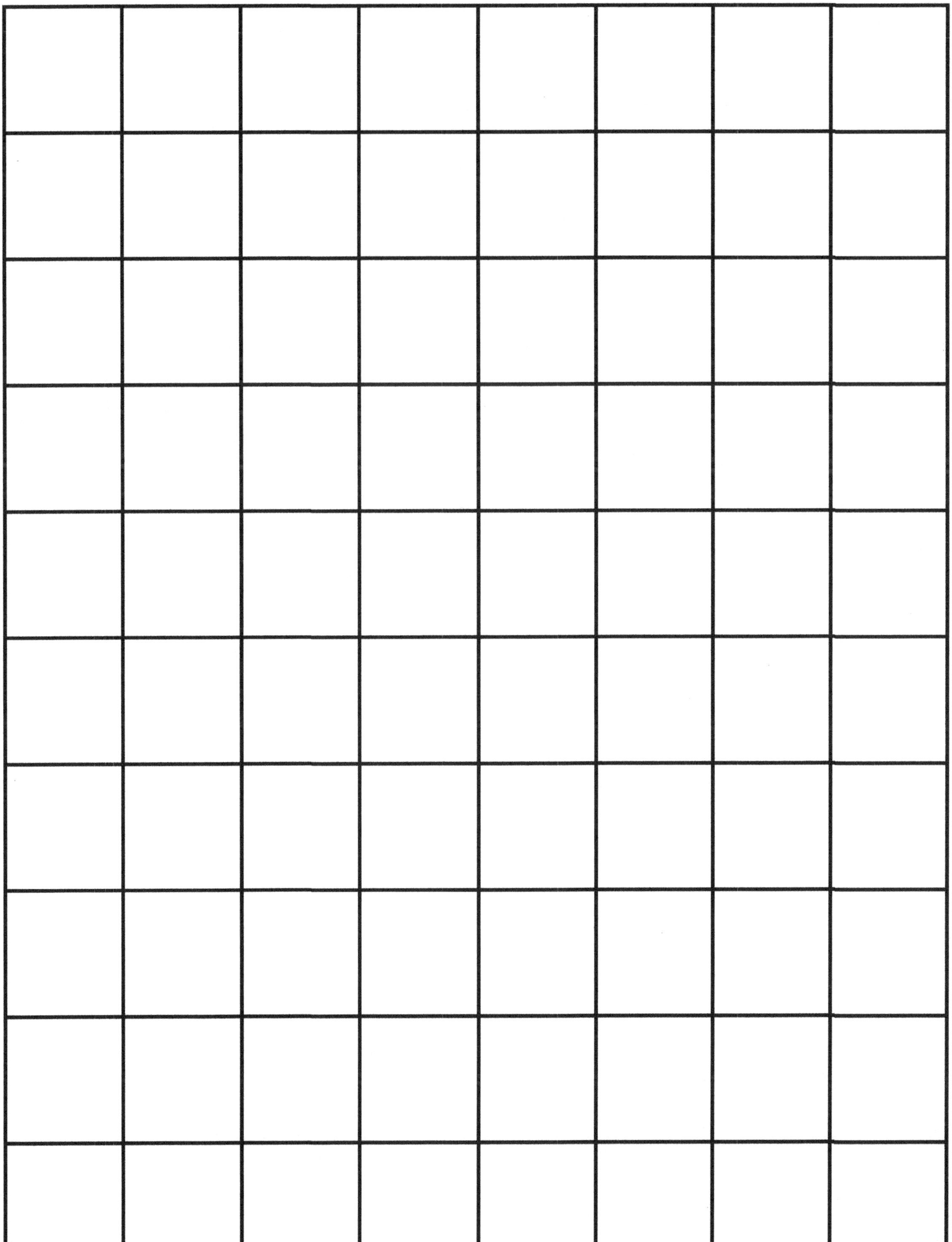

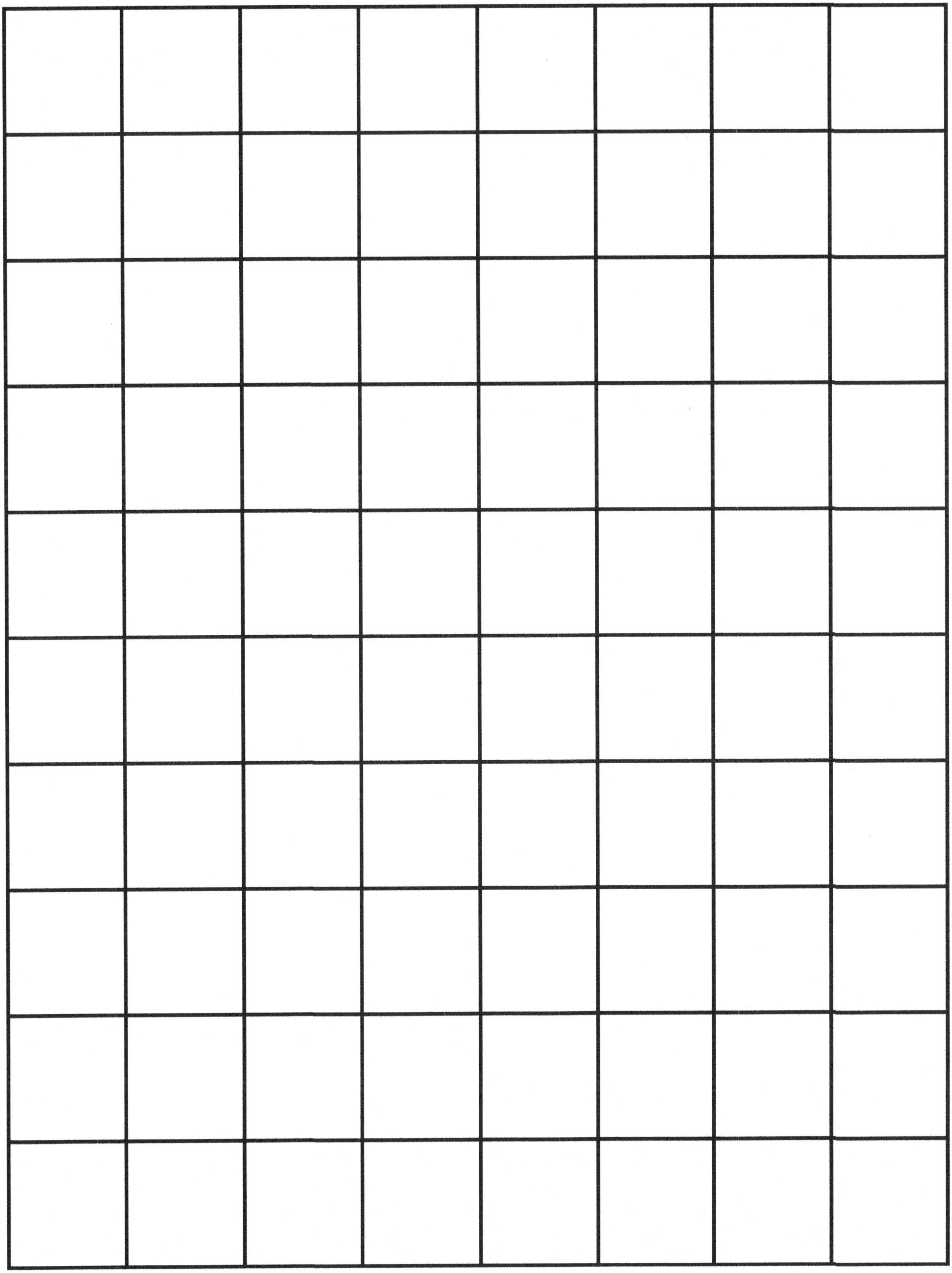

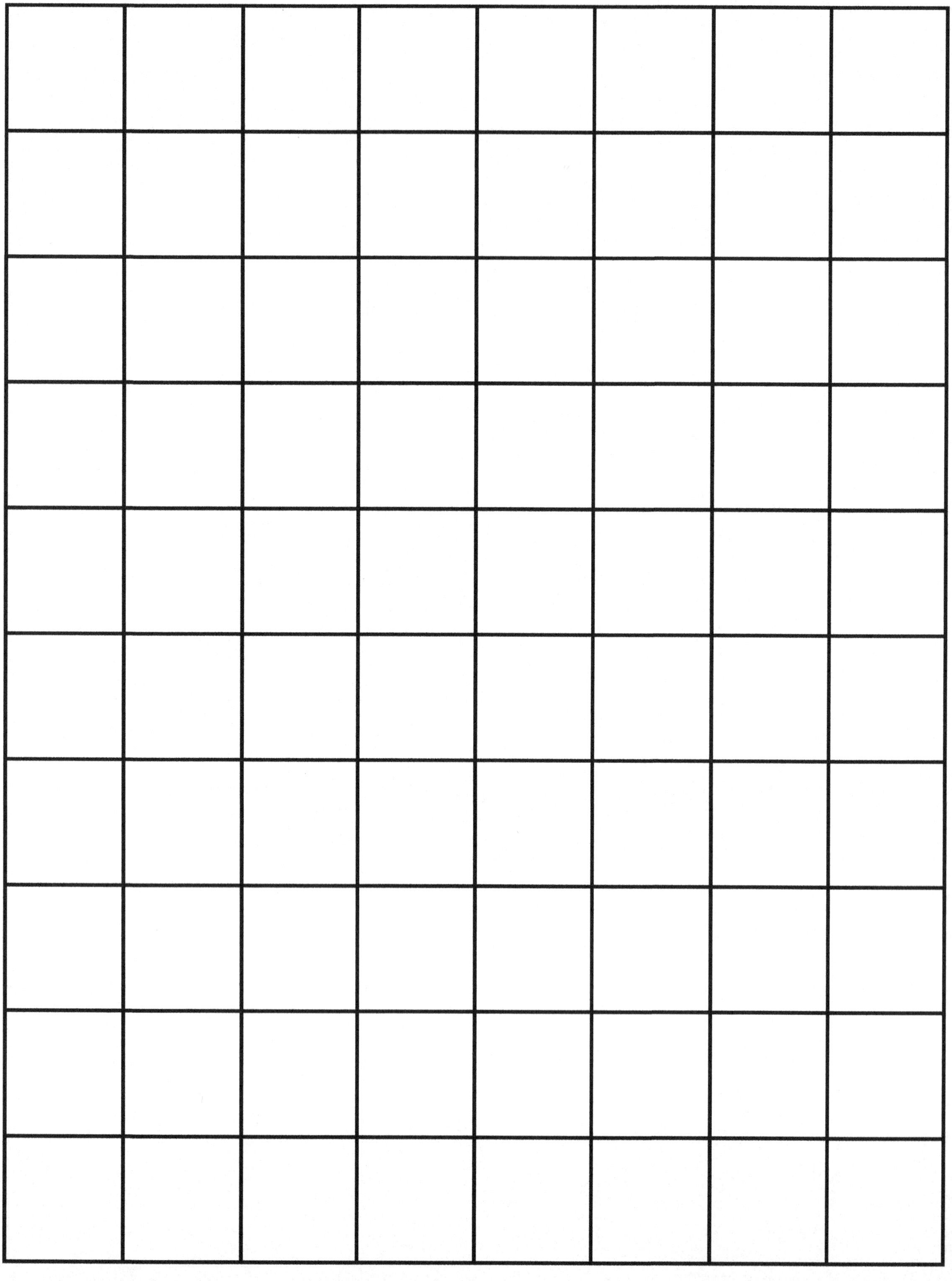

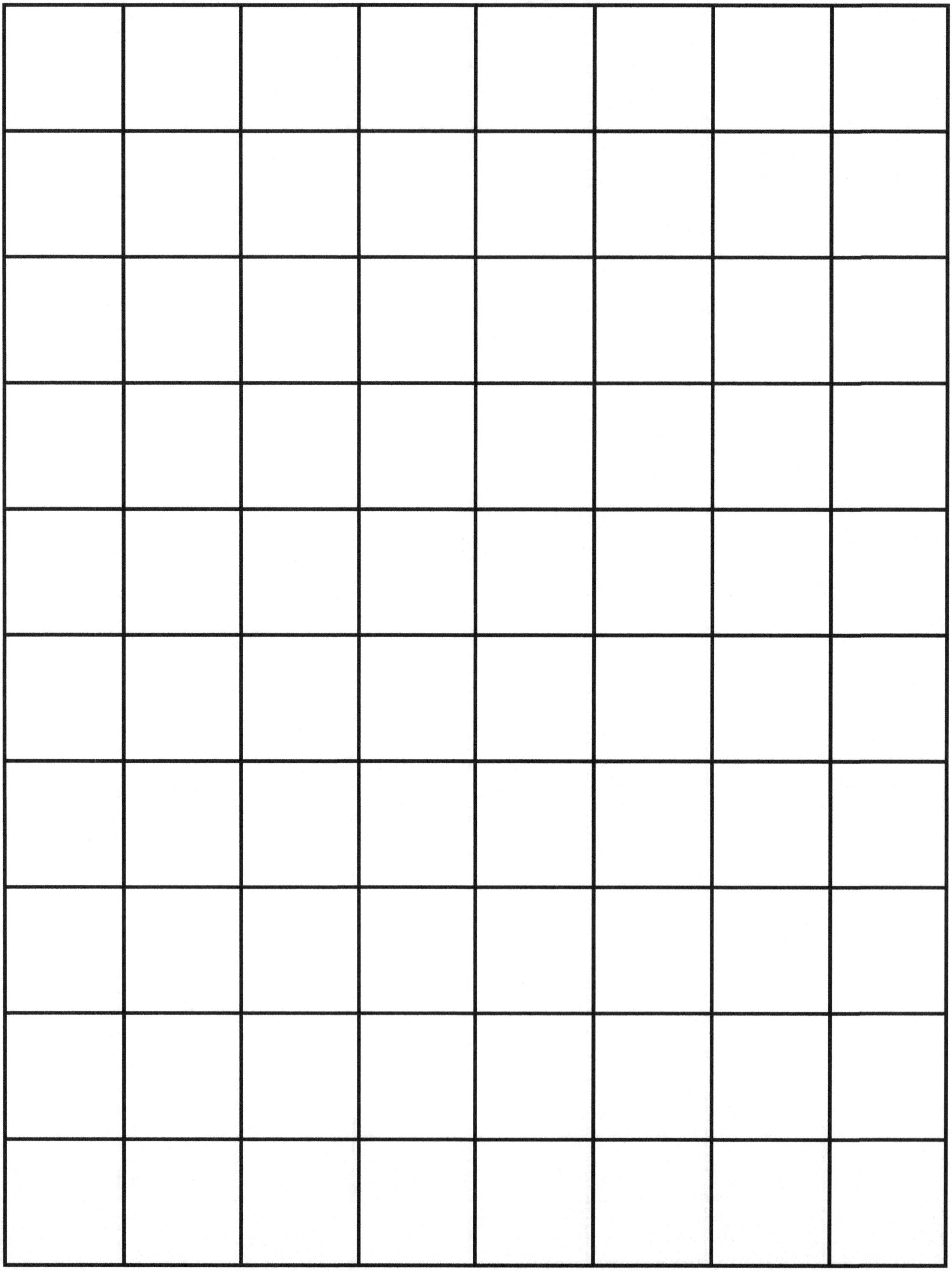

Made in the USA
Middletown, DE
01 March 2023

25984994R00057